Complete Electronic Media Guide

Complete Electronic Media Guide

Allen Lloyd

iUniverse, Inc.
New York Lincoln Shanghai

Complete Electronic Media Guide

iUniverse, Inc.

For information address:
iUniverse, Inc.
2021 Pine Lake Road, Suite 100
Lincoln, NE 68512
www.iuniverse.com

ISBN: 0-595-30939-9

Printed in the United States of America

Contents

1

BACKGROUND

INTRODUCTION

Storage and accessibility of information is critical to education, business, science, government and entertainment. The purpose of this document is to provide the reader with a brief overview of all electronic storage media available today and to demystify electronic storage technology.

In addition to media of the present, descriptions are included of older obsolete and "extinct" media to provide the reader with insights into how the technology has evolved. Whenever possible technical terms are explained.

TYPES OF STORAGE MEDIA

Electronic storage of information can be broken down into three broad categories: audio, images and data. Audio is the storage of sound, generally speech or music. Images are either still pictures or motion pictures (i.e. video). Data can be text, spreadsheets, computer programs, or any other type of information.

ANALOG VERSUS DIGITAL

Information can be recorded using either analog or digital technology. An analog signal is a variable electrical voltage that directly represents information. A digital signal consists of a stream of discrete symbols that indirectly represents the information. Audio and visual information is inherently analog (variable) in nature. The first devices to record sound and images were analog. They merely converted the natural analog information into artificial analog information. The photograph converts visual images into patterns of light and dark on a piece of paper. The old phonograph record converts vibrations of air (sound) into vibrations in a plastic groove. Both of these techniques are simply taking naturally occuring ana-log variations and imparting them to an artificial structure for preservation.

Digital recording is less intuitive, since the recorded signal bears absolutely no resemblence to the original "natural" information. In a digital system, a series of tokens or symbols is recorded. These symbols can represent analog data. The ana-log information is sampled (or measured) to convert it into digital data. These samples are numbers that describe the original information. When the data is "played back" it is simply converted back to analog where the original variations are recreated.

Digital data can contain much more than sampled analog information. It can also contain text, data files, computer programs, or any other kind of information.

THE GRAND CONVERGENCE

At one time, analog was the only game in towm. To store images there was pho-tography. To store sound there were tape recorders and phono-records. To store moving images there was motion picture film and analog videotape. To store text there was paper. Today digital technology can handle all of these types of infor-mation much more efficiently and conveniently. By using powerful new com-puter technology, all information can be handled as a file of digital data. There is really no need for analog storage or transmission any more. So why do we still have analog? The answer is multifaceted:

* INERTIA—Many people are quite happy with analog videotape, paper, film photography, and other analog recording devices. Also there is a signifigant investment tied up in analog media equipment and facilities. In broadcasting, for example, digital transmission is being phased in over a long time frame since there are so many analog-only receivers floating around and the conversion to digital is expensive on the transmit side.

* QUALITY—Some audio and video professionals are skeptical of digital tech-niques for recording sound and images. Hollywood still records most of its prod-uct on 35 millimeter motion picture film. The reason is that film still has the highest image resolution of any optical medium. Some audio "purists" still insist that vynl records sound better than CD's.

* COMFORT—Some folks simply prefer older analog technology out of habit. Scribbling notes on paper is second nature while fumbling with a PDA (personal digital assistant) is not. This is not as true for younger people who are much more open to new approaches.

* COST—An analog cassette recorder costs less than $25 and records up to 2 hours on one tape. A digital recorder with similar functionality is more expensive. A VHS VCR can be had for less than $60 at discount stores while the newer DV and Digital-8 video products are over $200. This is changing, especially for products such as hard drives and CD based media. CD-R blank disks are now cheaper than high-end cassette tapes and have much better quality.

For now, at least, there is still a healthy market for analog recording devices. There is no question that eventually all recording and transmission of information will be digital. This convergence is already beginning to occur.

There are several technologies that are facilitating this convergence of media. The Internet is at the top of the list, since it allows for e-mail (a substitute for the Post Office), MP3 and Real Audio (a substitute for radio), AVI and MPEG video (a virtual TV show), and of course newsgroups, web pages, and chat rooms which substitute for paper magazines The CD is also a powerful force for convergence, since it is much more than a music medium. Today CD's carry games, computer software, pictures, and even video. A 30 cent CD-R allows people to share incredible amounts of information and entertainment with each other, often cutting the mega-media corporations out of the loop. The DVD and DVD-R promise an even greater amount of this interpersonal proliferation of culture.

The "establishment" media corporations are not standing still. Television is converting to digital transmission. This will give them the ability to offer a multiplex package of channels with higher quality than the single analog channel they now have. Satellite digital radio is now available and terrestrial digital radio is just around the corner.

Cable TV, with its vast bandwidth, may become an all-purpose digital warehouse of material, mixing traditional real time "push" media such as TV channels with "on-demand" pull media.

MEDIA THEN AND NOW

Today's digital interactive media didn't spring up out of the ground. It is the result of steady progress in technologies such as electronics, optics, chemistry, and mechanics. The earliest audio and visual media were not electronic at all, but they were greatly enhanced with the introduction of electronic amplification, signal processing, and transmission.

During the period between 1860 and 1900 there was a flurry of new technologies that would forever change the way we relate to information and entertainment. The development of photography allowed for the storage and distribution of images. The moving picture democratized theatre, giving broad masses access to material that was once the sole province of the wealthy elite. The phonograph enabled millions to have access to the world's top musical talent on demand. New printing techniques enabled low cost distribution of vast quantities of data. This was all before electronics!

The key to making audio and visual media higher quality and more reliable was the development of the electronic amplifier. The electronic amplifier was a kind of "holy grail" which was sought by the telephone company to extend the reach of phones beyond the limits dictated by Ohm's law. Without an amplifier, the limit for connecting two phones was a couple of hundred miles and that was only possible with tree-trunk sized cables to prevent voltage drop.

In the early 1900's a man named Lee DeForest was tinkering with a Fleming valve (which was an Edison light bulb with a metal plate electrode inserted). DeForest tried sticking a little curly piece of wire between the filament and plate. He found that when a voltage is applied between the filament and this curious little wire, a much larger sympathetic voltage appeared at the plate of the device. He had found the Holy Grail! This device was called the triode since it had three elements and it allowed amplification of audio and radio frequency signals. This device allowed record players to have boomy room-filling sound instead of the frail tinny sound of the purely mechanical players. It also allowed for the era of sound movies. It allowed for coast-to-coast telephone service without thick wires. This one invention launched the age of electronic information.

Another invention that laid the foundation for today's hard drives and VCR's was magnetic recording. The earliest magnetic recorders used a spool of steel

wire. A tiny c-shaped electromagnet would magnetize the polycrystaline molecules of the wire, imparting an audio wave. On playback, the head was used as a
pickup device connected to a high gain amplifier. Obviously wire was not an
ideal medium since it was prone to metal fatigue and breaking. In World War II,
Nazi scientists improved the magentic recorder by substituting flexible plastic
tape for the wire. This tape was coated with iron oxide, which is ordinary rust!
This rust pigment was ideal for recording since it was very fine and thus had
many discrete tiny magnetic domains (which are like pixels in the photography
world). This results in good high frequency response and less noise. Another dis-
covery made by these scientists is that for best audio quality, it is necessary to
apply a bias to the tape to push it into the linear part of its transfer function.
Early machines simply added a DC to the recording head mixed with the audio.
This reduced distortion, but tended to roll off the bass frequencies. A better bias
was a low radio frequency (at around 70 KiloHertz). The AC-bias recorders produced shocking fidelity, making Hitler's tape recorded speeches sound live. At
the end of World War II, many of the German tape decks were confiscated and
some of them were used for American radio broadcasts!

The next quantum leap in recording came with the invention of video and video
recording. There were many brilliant pioneers who gave us modern video. The
most notable of these were Philo Farnsworth and Vladimir Zworkin, who
between them gave us the CRT-based scanning video systems in use today. Peter
Goldmark at CBS labs was another far-sighted innovator. He is credited with
many clever, if not commercially viable devices.

In the 1950's a young upstart company called Ampex came on the scene. This
comapny came out with the first practical video recorder. In order to make this a
reality there were enormous obstacles and challenges. Their biggest innovation
was the idea of "two dimensional" recording. Previous to this development, all
tape recorders simply dragged the tape across a stationary head. This was fine for
audio frequencies, but was totally inadequate for the radio frequencies found in
video. The Ampex idea was to have a spinning heads scanning across the tape to
combine traverse movement with the normal lateral movement. This greatly
increased the bandwidth, allowing for direct recording of video. The other major
innovation of Ampex was the idea of recording a modulated RF (radio frequency)
carrier on the tape instead of directly recording the baseband (raw) video. This
use of an FM (frequency modulated) carrier eliminated the need for expensive
wideband equalized linear head amplifiers.

In the 1950's computer tape was invented. This required the development of multi-track head technology and the use of pulse code modulation (PCM) to lay the data on the tape. A special form of PCM called NRZ (no return to zero) was invented to record data files. In the late 1950's and early 60's, error correction codes were developed that allowed much greater accuracy in digital data storage and transmission.

At around the same time, companies like IBM developed magnetic drums and crude disk drives which recorded circular tracks. This allowed for random access rather than sequential access to data files.

The first transistors that came out of Bell Labs in the late 40's and early 50's were germanium point-contact devices that were only good for low frequency audio applications. By the late 1950's silicon was being explored as a material for transistors. The silicon transistor had lower leakage than germanium units, making it more useful in digital switching systems. Using silicon, it was found that many transistors and complete circuits could be fabricated on the same substrate. This was the integrated circuit. In the 1960s digital integrated circuits became more and more complex. A new technique called MOS (metal oxide semiconductor) allowed very complex computer building blocks on one chip. In 1971 the Intel Corporation invented the single chip microprocessor. This ushered in the age of microcomputers.

In the 1960's the laser was invented. The laser or light amplification by stimulated emission of radiation is basically an optical oscillator. It generates light of one frequency as opposed to the many random frequencies found in ordinary light. This single frequency light is in a very thin beam. Early lasers were cumbersome tube devices that required high voltage power supplies. By the 1970's the solid state laser was developed. This was similar to an LED (light emitting diode). The solid state laser needed only a few milliwatts of power (less than a flashlight) and allowed for low cost, high density optical storage devices such as laser video disk players and the Compact Disc.

These are the key developments that laid the groundwork for contemporary electronic recording technology.

2

MECHANICAL PHONOGRAPH FORMATS

EDISON CYLINDER—This was the first instrument to record and play sound. It was composed of wax or metal foil. There were no electronic amplifiers, so a big horn was used to concentrate the sound. The tiny vibrations of the stylus were transmitted through the horn, "Recording" generally meant yelling at the top of one's lungs to modulate the groove. On playback a barely audible tinny version of the voice was heard. The cylinders held several minutes of sound.

78 RPM—This was the standard for "singles" in the early 1900s. This format used "fat" grooves compared with the later "microgroove" formats and was plagued with crackling noise and high distortion.

45 RPM—This was the later (1950's–1980's) single format developed by RCA. It had somewhat better sound (15 KHz) but still noisy. The "45" was the format used in jukeboxes and record changers. The "Top 40" radio format was originally based on the sales ofthese units.

33.33 RPM—This was the mainstay of the "recording" industry for multiple song "album" releases. Turntables ruled music for the better part of the twentieth century. The LP (Long Play micro-groove record) had very good sound quality and was capable of ultrasonic frequency response (which was used to piggyback FM carriers on top of the audio in the defunct quadraphonic disc format.) It was eclipsed by the CD which came out in the 1980s. (This speed was also used with 16 inch discs for "radio transcriptions" or prerecorded program distribution.) This format was developed by Peter Goldmark at CBS labs in the 1940's as a rival to RCA's 45 RPM single.

16.66 RPM—This is the forgotten speed. it was mainly used for special interest long-form programs such as reading of books and periodicals for the visually impaired, the ill-fated "Highway Hi-Fi" system of the 1950's, and the Seeburg "Lifestyle" long-play background music players. In the 1980's this format fell into obscurity, replaced (for spoken word book recordings) by special slow speed and multi-track cassette formats.

8.33 RPM—This speed was actually used for some long-form spoken word recordings. The wow and flutter must have been horrendous! Capacity was twice that of a comparably sized 16.66 platter. These disks are as rare as hen's teeth.

TED VIDEO DISC—This was a short-lived video disc invented by the German electronics firm Telefunken. TED discs were high speed phono records that held video information encoded mechanically. Each disc could only hold a few minutes and a multi-disc changer was needed for a full length program.
A good aspect of this format is that the disks could be pressed in mass production like audio records. This format never really caught on with the public, but it was a remarkable feat of German engineering!

CED VIDEO DISC—A strange variation on the phono-LP was the CED video disc format. This was released by RCA in the late 1970's and is formally called "capacitance electronic disc". This refers to the recording technique. In a normal phono record the information is imparted as "wiggly patterns" in the groove. These wiggles vibrate the stylus which converts the mechanical vibrations to an audio signal. A CED has grooves like an LP, but there is no wiggling of the stylus. Instead, the information is encoded as a series of conductive "pits" and "lands" which vary the capacitance of the metal-coated stylus. The subtle capacitance changes are used to modulate the frequency of a 915 MHz oscillator. The audio and video are recorded as FM carriers (from 0-10 MHz) and show up as "subcarriers" or sideband components of the 915 MHz carrier, which can be picked up and demodulated by conventional FM limiter/detectors. This technique was clever and practical and allowed cheap broadband playback without lasers or spinning heads. If the CD had not come out, this "groove capacitance" system might have been used to make hi-fi FM-encoded records. Who knows?

3

MAGNETIC AUDIO TAPE

REEL-TO-REEL—reel to reel audio tape used 1/4 inch tape running at several speeds. 30 IPS (inches per second) was used for the highest possible quality (24 KHz bandwidth and 70 dB signal to noise ratio). this was used by professional recording studios for mastering. 15 IPS was better than FM quality with 20 KHz response and 65 dB signal to noise ratio. It was also the speed used for high end pre-recorded tapes. 7.5 IPS would yield "FM quality" (a 15 KHz bandwidth and around 60 dB signal to noise ratio). 3.25 IPS would provide 10 KHz response and a 50 dB signal to noise ratio, good enough for talk or background music. It was a common sight in the early days of FM to see a "jockless" automated FM station using a rack of tape machines droning on for hours. Before satellite delivery of radio was common, tape was used for "canned" radio formats. Syndicators like Drake-Chenault and Bonneville shipped truckloads of FM-quality (7.5 IPS) reels to hundreds of affilliates. Reel to reel was the workhorse for "wholesale" audio for years. By the early 1980s cassette became the dominant "retail" audio recording medium.

CASSETTE—The reel to reel died as a consumer format in the 1980s. The Phillips Compact Cassette soon became the dominant recording medium. Originally the cassette was intended for voice applications such as dictation and answering machines. It was not taken seriously as an audio medium until the late 1970s when Japanese companies like Sony, Technics/Panasonic, Sharp, etc. started producing cassette decks with startling performance. Several issues were dealt with and systematically conquered. Frequency response of early units was on par with AM radio (7 KHz on a good day). This was mainly due to low density of particles on the tape and wide head-gaps. The Japanese companies weilded their R&D expertise and invented narrow-gap cassette heads. American tape companies started making much better tapes and invented Chromium-Dioxide ($CrO2$) and Metal particle coatings that allowed for better frequency response and a 2-7 dB drop in noise level. But even with all of these advances, cassettes were still below the magic "Hi-Fi" threshold of 60 dB signal to noise ratio. Chrome and Metal tapes had better sound and more high frequency headroom, but could only muster about 58 dB signal to noise. Enter Ray Dolby. His company, Dolby Laboratories developed a high frequency compression/expansion technique to "push" the noise away from the listener's perception threshold, allowing the "apparent" signal to noise to rise 10 dB with Dolby "B" coding, or as much as 20 dB with the later Dolby "C" system. Now cassettes could have over 70 dB signal to noise performance! This was the final nail in the coffin for consumer open reel. Throughout the 1980s the cassette improved. Companies like Nakamichi brought the cassette to un-heard-of levels of performance: flat response to 20 KHz, 75 dB signal to noise, and lower distortion thanks to dynamic biasing (Dolby HX Pro, etc). In 1985 the cassette ruled the audio world. Today the high-end cassette deck is becoming passe' since blank CD-R's cost less than blank good quality cassettes, and formats such as MD and MP3 have higher recording time in a smaller space.

EIGHT-TRACK TAPE—The 8 track tape was one of several cartridge endless-loop formats based on 1/4 inch tape. In the early 1960s several companies came out with such cartridges with slight differences in track layout and speed. Among these was the "Fidelipac" cartridge which was a 4-track cartridge that was used at 3.75 inches per second tape and played for up to 2 hours. The fidelity was mediocre and these cartridges were mainly used for in-store advertising and background music. Musical "appliances" such as church steeple bell players used endless loop fidelipac carts. A higher performance Fidelipac (running at 7.5 IPS) became popular with radio stations for recording "spot" announcements such as commercials and individual music selections. These became known as "carts" and were ideal for broadcasting since they automatically recycled themselves (no rewinding was necessary) and were easily cued. There were two consumer formats based on Fidelipac—the four track player came out in the mid-sixtees followed by the eight track (which could hold twice as much music). Of course, the eight track format had narrower tracks and thus an inferior signal to noise ratio compared with four track, but as with the Beta/VHS battle, the public voted for quantity over quality.

ELCASSETTES—The Elcassette was a short-lived format that used large cassettes running at twice the speed of normal audio cassettes (3.75 IPS). Introduced by Sony in the mid-70's, it offered better frequency response and lower wow and flutter compared to standard (Phillips) Compact Cassettes. The public response was a big yawn. The need for a "better" cassette was reduced as Compact Cassette performance greatly improved through better heads, transports, and noise reduction.

MICROTAPES—Microcassettes came out in the 1970s for lo-fi voice applications such as dictation, recording classroom lectures, answering machines, etc. In the early 1980s the Fisher company introduced a high performance.stereo microcassette deck (running at 1/2 the full size cassette speed). This deck had Dolby noise reduction and used special metal microcassettes. Although this unit had impressive performance, the public was not interested and the microcassette music format was quietly euthenized. Today microcassettes are being supplanted by small digital recorders based on "flash" digital memory technology. These have no moving parts and the contents can be "uploaded" to a PC.

DAT—the digital audio tape. It was also called "R-DAT" because it uses a rotary helical scan head. This was a format that was championed by Sony and released with much fanfare in 1987. The DAT was a small cassette (smaller than the analog cassette) that resembled a video cassette. It used direct digital sampling and recording (this was 5 years before MPEG) The introduction of DAT caused a stir in the recording industry. The RIAA at that time was warning that this could kill the recording industry since it would allow for totally flawless duplication of music. The industry responded by inventing SCMS or serial copy management system which limited the number of generations of digital copies that could be made. CBS Labs came out with a controversial idea to use a "notch" filter to delete a narrow band of audio frequencies in all prerecorded music. Any new high-performance recorder would be required by law to have a "CopyCode Scanner" that would look for this notch and refuse to record any material encoded with this characteristic. (Kind of a watermark). What they didn't take into account was the phase distortion that such a notch filter would introduce! Fortunately, the idea was never implemented. This became a moot issue since DAT was a non-starter in the consumer market. The only customers were radio stations and recording studios. In fact, the DAT was sometimes used as a 2-track mastering medium. The DAT has three sampling rates: 32 KHz, 44.1 KHz, and 48 KHz. These sampling rates have been adopted by many other media such as video tape recorder digital sound, satellite audio, Internet audio, and CODEC's (coder decoders) to digitize audio for fiber optic lines.

CODEC BASICS—So you might be wondering what is a CODEC? It is an acronym for CODER-DECODER. Basically, CODEC's are analog to digital and digital to analog converters. The CODEC takes a variable analog voltage (generally audio or video) and converts it to a string of numbers (called samples) that convey the instantaneous voltage. On playback or receive, it takes the string of numbers and reconstructs the original waveform (or at least an approximation of the original). Early CODEC's were "dumb" in that they simply converted the voltage to digital data or vice versa. To convert an analog voltage to digital, you must measure the voltage at regular intervals called samples. These samples are then recorded or transmitted to a decoder which takes the samples and converts them back to analog voltage. As a rule, the sampling rate must be at least twice as high as the highest frequency to be recorded (or transmitted).

Today's generation of CODEC's are "smart" in that they generally include a process to minimize the recorded/transmitted data stream. There are both lossless and lossy ways to reduce data. Lossless data reduction is used when the highest integrity is demanded. It is very similar to computer file compression such as ZIP, GZIP, ARJ, RAR, etc. Such systems convert repeating patterns into tokens, recording/transmitting only the token instead of the complete pattern. A lookup table is included so that the decoder can replace the tokens with the original patterns that they stand for. These systems are called lossless because no information is lost in a code-decode cycle.

Lossy compression is used for audio and video data because such data is extremely cumbersome and tends to hog a lot of bandwidth (or storage space) without drastic bit-rate reduction. Also audio and video waveforms contain a great deal of redundancy that does not contribute to the human perception of quality. It has been found that much of the audio or video wave can be simply discarded without any visible or audible degradation to the end user.

An example of an uncompressed medium is the original compact disc. The music CD audio is sampled at 44.1 KHz, Since there are two channels for stereo there are actually 88.2 KHz of samples. Since each sample is a 16 bit number, the bit rate is 1,411,200 bits per second or over 1.3 megabits per second! This requires a LOT of storage space and bandwidth! Fortunately, CD media is very cheap on a per-bit basis and there is adequate space. However, to transmit "CD quality" without compression, one would need at least 1.3 megabits per second. This would require at least a "T1" line (a special phone line with 1.544 megabit per second capacity) or 650 KHz of satellite bandwidth using a QPSK (quadrature phase shift keying which is typical for satellites). Both of these transmission systems are very expensive. This is an example of why lossy compression is favored today.

Lossy compression takes advantage of mathematical redundancy in a data stream like lossless compression, but also removes data that is regarded as inconsequential to human perception. In the 1980's and 1990's extensive research was done in the area of visual and audio perception. For visual data, resolution can be carefully rationed and applied to apects of the image that are most noticable. For audio, it has been discovered that when there are several sounds mixed together, some sounds will drown out others (a phenomenon called masking). By rationing the resolution, allocating more resolution to the dominant masking sound and less to the "drowned out" sound, signifigant reduction of data can be attained without a perceptible loss of quality. A $10,000 waveform analyzer can tell the difference, but your ears won't!

There are two main standards for compressed audio/visual data. JPEG (joint photographic experts group) is a standard for reducing still image data, while MPEG (motion picture experts group) is a set of standards for compressing video and audio data. There are many other compression schemes that are proprietary to one company or another (such as Sony's ATRAC, Real Audio, Visioneer MAX, Compuserv's GIF, etc) but JPEG and MPEG are the most widely used schemes for picture and sound data.

Within MPEG, there are many versions and layers. MPEG 1 was drafted in the early 90's. It provides a basic VHS-quality video signal and was used on the now-defunct Video-CD format. Within MPEG I there are three "layers" for audio reproduction. Layer 1 or MP1 is designed for moderate compression (384 kbps for CD stereo audio). It is rarely used these days. Layer 2 or MP2 is for moderate to high compression (192 kbps for near-CD quality) This is the most common audio format for satellite audio and is used for terrestrial digital TV sound and the DVD. Layer 3 or MP3 is optimized for low bit rates and is popular on the Internet (128 and 96 kbps are common for layer 3). Note that MP3 is NOT MPEG 3! It is MPEG 1, layer 3.

Most audio coding schemes split sound spectrum into bands (like a graphic equalizer). Because the human ear is not as sensitive to detail in the high treble and low bass regions, these bands are allocated less bits of resolution. The human ear is most sensitive to sound at around 800 hertz, so the bands between 100 and 3000 hertz receive the highest resolution. Although reducing bass resolution does not buy a great savings in coding bandwidth, reducing treble resolution can greatly reduce the bit rate without any perceptible drop in quality. The band from 10-20 khz can be allocated a small fraction of the resolution of the band from 0-10 KHz. In the analog world, the Dolby C noise reduction system actually reduces the audio detail (signal to noise performance) for sound above 10 KHz at high levels, but drastically increases signal to noise and reduces distortion for sound below 10 KHz, resulting in better overall quality. This tradeoff was made to prevent distortion and tape saturation at high levels but it is a kind of "data reduction" similar to the MPEG concept. It is important to note that today's MPEG-based audio is a very non-linear. In some MPEG audio, the transient distortion can be over fifty percent! This is "artistic distortion" that is designed to be mostly inaudible to the human ear, but it is a radical departure from the analog era when distortion above one percent was considered an atrocity. The presence of this"inoffensive" distortion is the reason that MPEG does not survive multiple decoding-encoding generations very well. This is also why some audio purists hate MPEG. But for the average end user, MPEG is a giant leap forward.

MPEG 2 is the video standard most frequently used today. It allows for everything from VHS quality to high definition and is found in most satellite dish systems, digital cable boxes, DVD players, and the new terrestrial digital broadcast tv.

MPEG video coding is much more complex than MPEG audio (layers I, II,and III). MPEG video transmits an entire image only when necessary and mostly transmits "update information" to transform the current image into the next image. (This is similar to delta modulation for audio signals where the difference between samples is transmitted instead of full new samples.) By only transmitting the minimum amount of data needed to "morph" the image, enormous bandwidth savings can be realized. Without MPEG, studio quality composite video sampled at 4fsc (4X the color subcarrier frequency) with 8 bits of resolution requires over 114 megabits per second of data. By using MPEG 2 video compression, bit rates as low as 5 megabits per second can be used for low action "talking head" channels, while 8 megabits per second is enough bandwidth for most movies and sports. As with MPEG audio, this is not a free lunch since there are sometimes "difficult" scenes that don't compress well. Scenes with lots of detail and movement sometimes overload the MPEG CODEC and result in "artifacts" such as tiling and "sawblading". Also movies that are "slow coded" by a computer or workstation rather than in real time (on the fly) tend to have fewer artifacts.

<u>DCC</u>—The DCC (also called S-DAT or stationary head DAT) was invented around 1991 by Phillips. DCC used an MPEG I, Layer I compression to bring 1.4 Mb/s CD data down to 384 Kb/s, This was applied to the tape via a multi-track thin film head (a technique borrowed from high capacity computer tape machines). DCC machines were backward compatible with the original analog cassettes and could play either format. The DCC failed in the market because there was little public interest in having an "improved" version of the cassette since most people were perfectly happy with the analog cassette. Purists shunned the DCC because it comitted the unpardonable sin of distorting the waveform. Also the release of DCC coincided with the release of Dolby S analog cassette machines, which stole some of DCC's thunder.

4

MAGNETIC VIDEO TAPE

VIDEO TAPE—Video tape was originally developed for broadcasters to store programs for transmission. The early machines were very large, cumbersome, and required "baby sitting" by a technician or engineer. Today video recorders can fit in your coat pocket and go anywhere. Uses of video production and recording have spread far beyond broadcasting. In fact, most video production today is for non-broadcast programming such as industrial training, corporate communication, and education.

REEL-TO-REEL VIDEOTAPE: The first practical video recorders used 2 inch wide tape and four heads mounted in a high speed rotary drum that would lay down FM-encoded video. These machines were called "Quad" because of the four head design. Before videotape came along, the only way to record a video signal was to aim a motion picture film camera at a TV screen. These were called "kinescopes." TV production in the 1940-1950 era was based in New York. Most shows were live for the eastern and central time zones and "kinescoped" for folks on the west coast. The Ampex Mark IV was the first video tape recorder good enough for the demands of broadcasting (there were some earlier attempts at video recording which produced very blurry images). The Mark IV was a monsterous "mainframe" unit that used several kilowatts and required frequent maintenance. (This was before transistors were common). In the 1960's, 1-inch formats were developed. Today 1-inch tape is still used in some TV production, but newer formats based on 1/4, 1/2 and 3/4 inch tape (such as Digital Betacam and DVCPRO) now dominate video production. Although the 2-inch format is totally obsolete, many TV stations, networks, and production companies still maintain thousands of these tapes. There is an effort now to transfer these tapes to more modern formats so that our "video legacy" will not be lost.

EIAJ—The electronic industry association of Japan, a coalition of Japanese electronics firms came out with a "standard" open reel video format in the late 1960's. The machines were quite bulky and this format didn't last long as it became apparent that a cartridge format would be necessary for a true mass market video recorder. The EIAJ machines, which used 1/2 inch reel to reel tape, were used for educational and business purposes, but were never a hit with the public.

U-MATIC was an early video cassette format that appeared around 1971. It used 3/4 inch tape and rotary heads to record video luma (brightness) and chroma (color) modulated to RF carriers. U-MATIC tapes were one hour in length. The tapes were LARGE and even clunkier than VHS. The main users of the format were in the industrial and educational arenas. U-Matic is still used. Many schools and libraries still have U-matic machines and U-Matic cartridges in storage. U-matic was extremely popular with TV news crews for news gathering. It freed local TV stations from having to use 16 millimeter film (no more "film at eleven"). Today there are millions of U-Matic cassettes sitting in storage of local and network news archives, which insures a market for U-Matic playback equipment will not disappear any time soon. In the 1980's Sony came out with U-Matic SP (superior performance). This increased the resolution and signal to noise performance. In the professional video arena, U-Matic was cannibalized by another Sony product, The Betacam SP format. Although U-Matic equipment is no longer being made, the U-Matic style cartridge with 3/4 inch tape is still used by video professionals, but with totally different recording formats that have better quality.

BETAMAX/BETA—This was the first widely distributed consumer video tape format. Invented by Sony in the mid 1970s, the "Betamax" machine became a popular consumer item. The main use of the new technology was, of course, taping TV shows. Sony was sued by the Disney studio on the grounds that the new machines were violating federal copyright protection.Sony prevailed in this case and the timeshifting era was officially underway. Technically, Beta is a 1/2 inch tape format that uses 2 rotary heads to lay two RF carriers on the tape, the luminance (lightness information) is recorded on a 4 MHz FM carrier, the chroma or color is recorded at around 688 KHz using quadrature (AM/PM) modulation. Audio is recorded using a stationary head as it is with audio cassettes. Beta was continuously improved with the addition of Hi-Fi (diplexed FM audio carriers), SuperBeta (increased resolution), BetaCam (a professional version of Beta for TV

news crews), and ED-Beta (extended definition) that surpassed video discs in resolution. In spite of all of these improvements, Beta failed as a consumer format,
losing to VHS, which had more recording time (but lower quality). BetaCam,a
broadcast-quality version of Beta aimed at the TV news gathering market, has
become an extremely sucessful format. Other formats that use the Beta tape cartridge include Betacam SP and Digital Betacam (both used in TV/Video production and archiving).

VHS or Video Home System was unveiled by JVC in 1976. It is similar to Beta
in that it uses two RF carriers recorded by rotary (helical) heads. The luminance
is a 4 MHz FM carrier with chroma at 629 KHz. (This is sometimes called "color
under".) Audio is recorded on a linear (stationary head) track. VHS, like
Beta,underwent a series of improvements such as Hi-Fi (FM audio tracks), HQ
(a boosting of contrast), and S-VHS or Super VHS, which increased bandwidth
for more resolution. Today VHS still rules video, but it is a lumbering dinosaur.
Video recording, like audio recording, is moving toward digital techniques. In
TV timeshifting, hard drive recorders are likely to take over, but there is a new
digital version of VHS under developement (called D-VHS, which records
directly from an MPEG digital packet stream.) In the movie market, DVDs are
already gaining ground from VHS. MPEG is here to stay, and MPEG-based systems such as DVD, Dish Player, and TiVo are the future of home video.
(NOTE—there are professional tape machines used by TV stations and networks
that use VHS-style tapes, but record with a much higher quality level such as the
Panasonic M2,D3, AND D5 formats.) 8 MILLIMETER is a format using a cassette roughly the size of an audio cassette tape. It was invented in the early 1980's.
Like VHS, it records RF carriers for luma, color, and hi-fi sound. It has about the
same quality as VHS. Hi-8, an improved version of 8 Millimeter, arrived in the
late 1980s. This includes digital PCM audio (8 bits, analog 2:1 companding, 15
KHz response). 8-MM has never been a serious threat to VHS, except in the
camcorder market. Hi-8 has become popular for some low-cost news gathering
and field acquisition. There is also Digital-8 which, as you might guess, records a
digital video signal on eight millimeter tape.

DVC OR DV is a relatively new digital VCR format introduced in the mid
1990s and supported by several Japanese electronics concerns. DV tape cassettes
are very small and use 5:1 lossy video compression. DV equipment is very high
priced and is not yet popular with consumers. DV pictures are DVD quality and

the sound is CD quality. This has made it attractive more for the low-end professional/industrial market rather than consumer use.

PROFESSIONAL STUDIO GRADE VIDEOTAPE FORMATS—There are several varieties of videotape used by the "pros" in TV and video production. D1 is a no-compromise digital studio format. It uses 3/4 inch tape cassettes and records uncompressed component video. This format came out in the mid-80s and has been supplanted by more efficient systems, but it is still used where cost is no object and top quality is demanded such as post-production mastering. D2 also uses 3/4 inch cassettes, but records digitized composite video (luma and color in one signal). It was popular in the early 90s but is out of favor today due to its high cost and composite (rather than component) recording system. D3 is a 1/2 inch version of D2. D5 is Panasonic's component digital video format (similar to D1, but with 1/2 inch metal tape). Betacam SP is a very popular analog component format that uses "Beta style" 1/2 inch tapes. It is the most popular professional format for news gathering, post production, and even on-air playback. Betacam SP is slowly being replaced by digital formats such as DVCPRO (a Panasonic digital format), DVCAM (a Sony digital format), Betacam SX, a digital version of Betacam optimized for news gathering, and Digital Betacam, another digital version of Betacam designed for in-studio production and on-air playback.

It is worth noting that day to day on-air playback is now shifting away from all tape formats and toward hard-drive based video "servers" which, combined with automation software and wide area networks will eventually make all television production and transmission 100 percent digital.

5

COMPUTER TAPE FORMATS

COMPUTER TAPE FORMATS—Tape has been used for storing digital computer files since the 1950's. In the early days, tape was used as the main storage medium for programs and data. Today it is used mainly for backup of hard disk drives and for near-line (as opposed to on-line) mass storage. Tape is a serial access medium, so finding a piece of data requires scanning through the tape, which can be time consuming. Disk drives have a FAT (file allocation table) and are random access, allowing nearly instant access to any piece of data. Although tapes are slower than disks,they are much more robust and stable, making them an excellent medium for backup and general long term storage of files.

TAPE ARCHIVING—When files are stored on tape, they are often grouped together into one large continuous archive file. The Unix operating system has a popular utility called TAR (for Tape ARchive) which is often used for writing a group of files to a tape for backup or distribution purposes. The TAR file is read as one file and unbundled after is read to recover the original files. Often TAR files are run through a compression program such as GNU Zip (or GZIP) which is a "lossless" compression scheme that minimizes the size of the file through a 100 percent reversible process. Other programs implement the grouping and compression at the same time. Note that such archive files can be stored to any medium, not just tape. Other popular archive formats include ZIP, CAB, ARJ, and RAR.

NEAR-LINE VERSUS ON-LINE—An on-line medium is one whose data is available nearly instantly when demanded. Magnetic disk drives (both floppy and hard) as well as solid state storage such as RAM and ROM are examples of on-line media. Near-line media are similar to on-line, except that there is some delay, generally a few minutes, between a request for data and its delivery. Tape drives can be set up so that their contents are available on demand, but this generally involves "mounting" the tape on an available drive and then searching through the tape for the desired contents, both of which are time consuming compared to random access media. Optical media such as CD-ROM and DVD-ROM are "borderline" cases since their transfer rates are much slower than hard drives, but they are random acess devices.

OFFLINE STORAGE—An offline medium is one that is not available without human operator intervention. Typical uses for this are archiving and backup of data such as financial records and data retention required by the Government. The most common medium for this is computer tape such as IBM 3590/3580, Ampex D2/DD-2, DLT, or similar formats. Some organizations may use a CD or DVD based format, but at present, tape is the most popular solution since it still has the lowest cost per bit for storing large volumes of archive data and has a faster read/write rate. Some organizations with an old archive base may use the old IBM 9-track format or even microfilm. Surprisingly, paper is still a viable archive format. Today's acid-free paper has a life span measured in centuries. One can go to a landfill and dig up newspapers thrown away decades ago and still read them (and smell them!) Of course, there is no way to electronically search paper to find the material you are looking for and the density of data is low. But it is true that a computer printout of a file may outlive the disk or tape it came from!

AUTOMATIC TAPE LIBRARIES—For large mainframe and minicomputer centers that need to store vast amounts of data, high capacity tape is preferred to hard disk storage because it offers enormous capacity and a lower cost-per-bit as well as better long term stability. In the early days, tapes were plucked from a shelf and manually installed in available tape drives by a data center employee. This manual system was slow, expensive, and prone to error. Manual loading of tapes is still used for some backup and inter-facility data transfer, but for near-line general access, the automatic tape library was created. The automatic tape library is combines one or several tape drives with a robot that grabs tapes from a shelf and loads them into the appropriate drive where the contents can be down-loaded or new contents uploaded. Tape libraries range in capacity from around one terabyte (1000 gigabytes) to one petabyte (1000 terabytes). ATL's are some-times called silos. Each tape is identified by a bar code label.

DATA TAPE FUNDAMENTALS—When data is transmitted or stored, it can-not be simply "dumped" in a contiguous stream. In order to insure the reliability of the data and allow easy synchronization of the playback/receive side, it must be transmitted as "blocks" or "packets". The word block is used to describe chunks of data stored on tapes and disks, the word packet refers to chunks of data that are transmitted over communication links. Each block or packet has identification information and error correction data.

In a data storage system, error correction is necessary to insure the integrity of the data read from the medium. Generally this involves adding redundant information that can be used to recreate the packet if it is found to be defective. A mathematical test is applied to each block. If the block fails the test, the redundant data imbedded in the block is used to recreate the missing or bad data. If the "corrected" block then fails the test, the data is lost and cannot be recovered. In practice, uncorrectable errors are extremely rare, especially if there are multiple layers of redundancy and error correction. This type of error correction is called forward error correction and is used for data recorded on all disk and tape formats as well as one-way transmission links such as satellites. The mathematical test applied to each block of data is called CRC or cyclical redundancy check.

Tape systems record data as a series of blocks. Typically the block size is 512 bytes but may widely vary.

The problem with computer data storage is that the "native" format of computer data is DC voltage levels. Within a computer there are collections of transistors (on chips) that store a "one" as +5 volts and a zero as ground level (zero volts). This works fine for the TTL (transistor transistor logic) or CMOS (complementary metal oxide semiconductor) chips, but one cannot store level data on a magnetic tape directly. The head of a tape recorder is really one side of a transformer. On record, the head is a primary winding while on playback the head acts as a secondary winding. One problem with transformers is that they can not pass DC (direct current) which means they can not transfer continuous logic levels. (Magnetic disk heads work on the same principle). To properly exploit tape recording for data, the data must be modulated to a carrier wave or a pulse stream. There are several ways to do this, but all of the techniques encode the data by manipulating the timing of the waveform (or pulse train) to impart the ones and zeroes. The coding schemes for tape and disk data are designed to minimize DC offset, which is a bias toward negative or positive. In other words, the average voltage of the pulse stream should be zero.

REEL TO REEL COMPUTER TAPE—This tape format was the mainstay of data processing for many years. This was a 1/2 inch iron-oxide tape similar to VHS tape (without the shell). The tape was on a 600-3600 foot reel. (This was the type of computer tape shown in spinning in many Hollywood movies of the 1960's and 1970's.) These tapes hold nine parallel tracks of data. One track was for synchonization and each of the remaining eight tracks carried one bit of data so the tape recorded one byte at a time in parallel bits. The data was recorded using pulse train carrier techniques. These old tapes could only hold about 200 megabytes and are now hopelessly obsolete. In spite of this many companies still maintain thousands of these old tapes. There is a danger that the tapes themselves will degrade and become unreadable. Also it is getting harder to find playback hardware for these tapes. This is similar to the situation with old obsolete video formats such as Quad. There is a "race against time" to convert these tapes to newer formats while playback equipment is still available.

3480, 3490, 3580, AND 3590—These are IBM 1/2 inch cartridge tape formats that use chrome and metal formulation (instead of iron oxide) for higher bit density than previous linear tape formats (such as the old 9-track format). Another improvement is the so-called "serpentine" method where the machine records parts of the data in one direction and other parts in the other direction. This is similar to the "auto-reverse" feature on audio cassettes. These tapes are called "square tape" since the reel is enclosed in a square plastic cartridge instead of being an "open" reel. Many IBM/mainframe/COBOL shops employ these tapes for recording mission-critical corporate data. These tapes are extremely robust and can last for decades in proper storage conditions. These tape carts are also used for transferring data between mainframe/mini facilities and for software distribution. Capacity is in the 1-100 gigabyte range. Often these tapes are coupled with robotic automatic changers in large data centers.

DLT/SDLT—The Digital Linear Tape format is another 1/2 inch cartridge format deveoped by Digital Equipment Corp. It is similar to the IBM format, but optimized for DEC minicomputers (VAX, PDP-X, etc). Today DLT is still used (mainly for hard drive backup). Capacities of 100 GB or more are common (with compression). DLT is often used in robotic tape changers or "jukeboxes" for nearly instant access. SDLT is Super Digital Linear Tape, a higher capacity version of DLT.

QIC—The quarter inch cartridge was developed by 3M in the early 1970s. It is similar to an audio cassette, but the tape is twice as wide and there are multiple tracks on the tape to "inverse multiplex" the data. QIC never caught on in the mainframe world due to its rather low capacity and slow speed. It has mainly been used to backup data for workstations and small minicomputer systems.

TRAVAN—The Travan format is a new and improved version of QIC. The tape is slightly wider and the transport is faster and more robust than the QIC drives. Still it is limited to low end systems. Travan and QIC have been eclipsed by high density linear and helical formats.

DDS—The Digital Data Storage cassette uses DAT (4mm) format tape, but stores raw computer data rather than digitized audio. It uses helical scanning similar to a VCR instead of multiple parallel tracks employed by the IBM and DLT formats. Helical scanning is somewhat rough on the tape, which can lead to a shorter lifespan than for linear stationary head recording. Also helical heads require frequent cleaning and replacement compared to stationary heads. In spite of these problems, it has an impressive capacity for its size and price (over 40 GB). DDS is mainly used PC's and small minicomputers for hard drive backup. DDS is considered too "delicate" to be a workhorse for mainframes.

MAMMOTH—This is a digital data format similar to DDS, but it uses eight millimeter tape that is very similar to the eight millimeter video cassette. It too uses helical scanning and has the same head wear and tape wear issues of DDS. Mammoth is a proprietary format of the Exabyte company. Exabyte was the first company to exploit consumer helical scan video technology for storing computer data. In order to meet the stringent reliability requirements for storing mission critical data, Exabyte greatly improved and refined the eight millimeter VCR transport, minimizing friction and adding a "read after write" feature that checks the integrity of newly recorded data, rewriting it if errors are detected. Mammoth tapes are popular in the Unix world for backup, near line storage,and transferring data between machines.

AIT—or "Advanced Intelligent Tape" is Sony's answer to Mammoth. Like Mammoth, it uses a modified 8mm video cassette. AIT tapes can hold over 100 GB.

DTF—The Digital Tape Format is a special version of Sony's popular Digital Betacam broadcast VTR tape that is optimized for long duration archiving of high-density data (over 100 GB per cartridge). Often these tapes are stored in robotic "jukebox" machines for near-online access.

LTO—LTO or linear tape open is a newer format implemented by several companies. It is called "open" because it is not proprietary to one company and, like the DVD or CD, can be licensed to any manufacturer (if they meet the stringent quality standards). IBM and HP are the biggest backers of LTO at present. LTO, as the name suggests, has a linear (non-scanning) stationary head using advanced thin film multiple parallel track technology (sometimes called inverse multiplexing). LTO implements many advanced features for quality monitoring and management for the absolute highest overall data safety.

VTR-BASED DATA TAPE FORMATS—There are several other computer tape formats based on VCR helical scan technology. D1-D6 professional video cassette formats are sometimes used for very high capacity mainframe data storage. Capacities of over 200 GB per tape can be realized. This is the domain of well funded corporate information centers (the machines cost around $100,000).These tapes can be put in a vault for backup or offline storage or held in a robotic tape changer for near-online access.

6

WRITABLE DISKS

<u>CD-RW</u> The CD-RW is an optical storage device that is coated with a special layer of crystalline material that, when excited by a laser, changes reflectivity and thus can store information. Unlike a CD-R, where a laser is used to burn little pits in a dye layer of the medium to encode the information, CD-RW uses a laser to change the reflectivity of a crystalline substrate. This process is reversable, permitting the disc to be erased and re-written thousands of times over (like magnetic tape and floppy discs). Like their CD-R cousins, CD-RW has around 650 MB capacity. To change information on CD-RW, the session must be completely erased and re-written. This makes it less flexible than a hard drive, where files can be erased and re-written individually.

<u>DVD-RW</u> the DVD-RW is basically the same as CD-RW, but it has over 5 GB instead of 650 MB of capacity. DVD-RW discs can not be played on CD-only players or drives. DVD-RW now has a strong competitor in DVD+RW (see DVD+RW section) and no longer looks inevitable. DVD-RW, like DVD-R, can be played on a consumer DVD movie player, but only after the disc is "finalized" and recorded in a special compatible mode. This makes it somewhat unattractive for real time "on the fly" recording such as camcorders.

<u>DVD-RAM</u> the DVD-RAM is similar to DVD-RW, but it can be selectively erased and re-written. On a normal CD/DVD-RW, you must erase the entire disk to write new material. The DVD-RAM is somewhat analogous to the Sony MiniDisc, the Iomega Zip format,and the Imation Superdisk. DVD-RAM, unlike other DVD-based systems, is encased in a protective caddy. This keeps dirt and fingerprints away from the media surface, but makes it totally incompatible with any other DVD format.

MD or MiniDisc—was invented by Sony in the early 1990s. Like DCC, it used an MPEG-like compression (ATRAC).The MD has enjoyed some success. It is based on magneto-optical recording that allows disks to be erased and re-used like magnetic tapes. MD's have become popular in radio stations, since they have better than FM (but not quite CD) quality and have features like track editing and cueing that make them ideal for recording spots and music tracks. MD's have fared less well with the public at large, especially with the availability of low-cost CD recorders and CD-R media. Also the MP3 format has eclipsed MD for convenience and portablility. This is unfortunate since MD has generally better quality than MP3 and is very easy to use. MDLP is a new MiniDisc long-play format. The original MiniDisc records a data stream of 292 kilobits per second for stereo or 146 kilobits per second mono. The new version of ATRAC used in MDLP recording uses 132 kbps and 66 kbps and is a much more aggressive coding algorithm similar to MP3. This allows five hours of audio on one MiniDisc.

DVD+RW (pronouned DVD PLUS RW as opposed to DVD DASH RW) Not to be confused with DVD-RW, DVD+RW is more like the MiniDisc or DVD-RAM than the CD-RW or DVD-RW. One of the problems with DVD-RW is that, like CD-RW and CD-R, it must be "finalized" before it can be read by a stand-alone reader or player. If you wish to modify the contents after the "finalizing" or "session closing" stage, you must erase the whole thing and start all over! This makes it inconvenient for impatient consumers. The designers of DVD+RW wanted a truly random access medium which, like magnetic tape, can be played on a stand-alone player without the cumbersome "session closing" stage required by most CD and DVD recording media. Another design goal was to make allow for easy re-arrangement of the contents (like the MiniDisc, where tracks can be added, deleted, or re-ordered "on the fly"). DVD+RW video discs can be played on conventional stand-alone DVD movie players without finalizing. For compuer storage, DVD+RW discs are analogous to the Iomega Zip format or DVD-RAM, providing "drive letter" access and allowing for totally random access to files like a hard drive. DVD-RW and CD-RW are all-or-nothing propositions, requiring a complete session erasure for modification of contents.

FLOPPY DISKS are small low-capacity computer file storage devices. The old 5.25 inch floppies are obsolete (they could only hold 360K bytes!) The 3.5 inch floppy diskc are still used, but they can only hold around 1.4 megabytes and are being replaced by higher capacity formats. These are sometimes called "diskettes". Floppy disks are round pieces of magnetic tape (iron oxide on a flexible plastic base). When the disk is in use, it is spun at a high speed (hundreds of RPM) which causes it stiffen through centrifugal force. In this state, a head can ride above the disk surface and lay down circular tracks on the surface. Like magnetic tape, floppy disks cannot record a DC constant voltage (zero hertz) so level-based coding of data used in semiconductors will not work. Instead, the data must be modulated to a square wave train of pulses. These pulses form a carrier, which can be modulated to carry the data stream. This concept is similar to the VCR recording of video information, in which an FM carrier modulated with the video signal is stored on the tape. The main difference is that the VCR video carrier is at several megahertz while the "floppy carrier" is less than one megahertz and is a square wave. The data is modulated to the carrier using a MFM (modified frequency modulation) where the position of the edges of the pulses imparts the data. This pulse modulation is similar to the RLL (run length limited) used on hard disks, NRZ (no return to zero) used for many computer tapes, and EFM (eight to fourteen modulation) used on CD media. Floppies are practically obsolete with a capacity of less than 2 MB, slow access speed, and vulnerability to damage from magnetic fields. The replacement for the floppy disk will likely be some form of CD or DVD.

ZIP DISC—is a proprietary format of the Iomega Corporation. Zip discs hold 100 megabytes or more. Zip discs are based on a selectively erasable and rewritable technology similar to MiniDisk and DVD-RAM. They are in a protective cartridge. Due to the relatively high cost of Zips and the declining cost of writable CD/DVD media, it has been losing ground in the market. Also the proprietary nature of this format is a source of worry for archivists. Still, it has found a niche for short term storage and data interchange. Iomega hopes that Zip will replace the 3.5 inch floppy for general purpose portable storage. Zips can also be used in non-PC applications, such as audio recording and camera image storage. (Jaz is a similar Iomega format that holds 1-2 GB). Both Zip and Jaz provide truly random read and write capability. Which makes them more user-friendly than CD media.

SUPERDISK—Superdisk is a proprietary format of Imation. Superdisk has around 120 megabytes of capacity and is the same size as a 3.5 inch floppy. In fact, Superdisk drives will accept the old 3.5 inch format as well as the Superdisk. Superdisk can be used in non-PC devices for holding audio tracks and pictures. Like it's cousin, the Zip disk, the Superdisk has not caught on well in the market.

Many corporate and individual users do not want to make a signifigant invest-ment in a format that is proprietary to one company. Also the declining price of recordable and erasable CD's and DVD's has hurt this format. One plus of this format is that it provides "drive letter access" which allows random reading and writing similar to a hard disk, unlike CD-R and CD-RW which must be "final-ized" and have limited random access.

HARD DRIVES—The computer hard drive is a sealed non-removable high capacity magnetic disk system that can store many gigabytes of data for immedi-ate (online) retrieval and manipulation. Hard disks store a "map" of all available files called the FAT or file allocation table. This allows the hard drive to instantly retrieve a desired file anywhere on its surface. Hard drives are very fast, measured in megabytes per second whereas other media such as CD-ROM are measured in kilobytes per second. Hard drives are second only to solid state (RAM, ROM, etc) systems for speed. At some point in the future desktop computers may become diskless, using only solid state for mass storage, but that seems highly unlikely any time soon since the price-per-bit of hard drives keeps falling.

Data is recorded and played back on a hard drive by a magnetic head similar to a tape head. Unlike the tape head, the hard drive head does not touch the media surface. Instead, it "flies" slightly above the media surface. Data is written and read by this head. As with any magnetic medium, hard disks cannot record level data (DC voltages) so the data must be recorded by modulating a carrier. This carrier is a train of pulses whose positions (relative to each other) imparts the data. Most of today's hard drives use a form of pulse position coding called RLL or run length limited. This is a form of FM where there is are maximum and minimum intervals between pulses. Hard drive controllers communicate with the computer, translating the computer's native data format to this special pulse for-mat used on the hard disk. The RLL scheme is similar to EFM (eight to fourteen modulation) employed on CD's. Most of today's hard drives have all the com-puter-to-drive format conversion build directly into the drive itself. This often referred to as IDE for Integrated Drive Electronics. Early hard drives put a bur-den on the host computer to do format conversion. Today the converters are in firmware (software permanently wired into silicon chips) which is part of the drive's controller. This IDE controller is usually a card plugged into the main board of the computer, with 40-wire ribbon cable (the IDE bus) connected to the physical disk drive itself. A competing bus technique, called SCSI (pronounced "scuzzy", a bad acronym for Small Computer Systems Interface) uses a similar controller card plugged into the main board of the computer. While IDE uses a flat unsheilded ribbon cable and is limited to internal (within the computer body) interfacing, SCSI cables are external and plug into the back of the computer to connect external devices. SCSI cable are thick, with a grounded shield surrounding the main conductors to keep out stray interference. Unlike IDE, which is predominantly a disk drive (and CD/DVD) format, SCSI can be connected to a wide variety of peripherals.

HARD DRIVE RECORDERS—There are two types of HD recorders, the dedicated or "hard wired" devices and those based on PC software. The latter are mainly used for radio stations (as part of a larger studio automation system) where spot announcements, voice tracking messages, news, and music tracks can be organized for unattended operation. Hard-wired dedicated HD machines come in several varieties. The TiVo is a video HD recorder that is a box with a hard drive and an MPEG CODEC (coder/decoder) and operates as a stand-alone device similar to a VCR. With the cost of HD's plummeting, we will probably find HD's popping up in many consumer devices such as camcorders, and stereo components.

It is actually possible to use a hard drive as an analog device. In the late 1960's the Ampex company invented an analog hard drive device for capturing a few minutes of video for instant replay. The device was used mainly for sports broadcasts and could replay in slow motion and could provide still frames. This used an FM carrier modulated by analog video instead of a bit stream.

7

OPTICAL DISKS

PIONEER LASER DISK—this was one of the first optical disc formats released to the general public. The Pioneer LD used analog (FM) coding of audio and video and provided the home viewer with true studio quality. Optical disks, like magnetic media, cannot record direct current. The information is imparted to the media by burning pits in the surface. These pits can carry pulse coded analog signals or digital signals. To carry an analog signal, the length of the pit is varied in accordance to the analog waveform. The Pioneer LD records several FM carriers on the disk. An 8 MHz FM carrier carries a composite color NTSC signal. The stereo sound channels are carried at 2.3 and 2.8 MHz in wideband FM. These analog FM channels are nearly CD quality with very wide deviation. In the early nineties, digital audio was added to the LD format for true CD quality. (These disks also had analog sound for backward compatibility). In the mid nineties, Dolby AC-3 surround sound was added to the LD format. Unfortunately by that time the writing was on the wall that LD would soon be replaced by the DVD, which is totally digital.

The LD format, while never challenging tape formats such as VHS with the broad masses, became a favorite of "video snobs" and movie collectors. Many movies were released in this format and it became popular as a home theatre format to use in conjunction with big screen TVs and fancy surround sound stereos. Although new machines are no longer being made, many collectors have amassed hundreds of titles and are not about to abandon it. One of the advantages of LD is that it uses no video compression (such as MPEG) and provides true uncompromised studio quality. DVD, by contrast, uses drastic bit-reduction with its MPEG codec. Theoretically this should be transparent, but some DVD's will show artifacts on fast changing, complex scenes.

The Pioneer LD format was able to produce very high quality freeze frames. One of the uses of the format was for storing collections of still images (a video version of microfilm).

COMPACT DISC—this is the most well known and used optical disc medium. The original CD was designed for multi-song stereo music program of 74 minutes or less. It uses straight 16 bit PCM coding and no data reduction (MPEG, ATRAC, ADPCM, etc). The audio data is stored on the disc using FEC (forward error correction) which records redundant data to correct any errors that might otherwise lead to "dropouts" or static-like noises. All CD's, (both music and CD-ROM) use a DC-free pulse position modulation technique called EFM or eight to fourteen modulation. Briefly, this converts eight bit "logical" bytes into fourteen bit "channel" bytes. These channel bytes follow certain rules regarding minimum time and maximum time between ones (to minimize confusion to the decoder). These 14-bit symbols can be stacked one after another without too many back-to-back zeroes or ones (which would cause a loss of synchronization). The logical (8-bit) bytes are converted to channel (14-bit) bytes (and vice versa) through a look-up table. This type of modulation, coupled with the Reed-Solomon error compensation, results in very high reliability.

Since CD's are devoid of compression such as MPEG, it can be described as a true studio quality medium. Because of a fortunate historical accident, CD's came out several years before data reduction schemes were available, thus it was established as the absolute "gold standard" of audio. Another fortunate fact is that the CD's architecture is totally open and unencrypted (raw PCM data like a computer WAV file) making it very easy to "rip" audio data for importing to a computer or other device.

It's fortunate for owners of CD's, not for sellers, who have been gnashing their teeth due to the ease of replicating CD's and transmitting CD data via the Internet. In 1982, when the "Red Book" audio CD was standardized, no one considered the possibility that the public will be able to buy blank CD's and make perfect "bastard" copies and have "any to any" connectivity on their home computer via a high speed network to anyone else in the civilized world. As it stands now, the business of selling pre-recorded CD's is facing financial calamity due to these "software security" problems. It is clear that the designers of the DVD format learned from the painful lessons of the music industry and thus incorporated such security measures as CSS (content scrambling system) to make it difficult to "rip" DVD data (import it into a PC) and "country codes" to regionalize content. Eventually all entertainment will be released on DVD and music CD's will likely be phased out in favor of "music DVD's" which will have the added security measures of DVD. In the meantime CD makers have a few tricks up their sleeve:

INTENTIONAL ERRORS—One technique used to foil the casual "rippers" is to take advantage of a glitch of most PC CD drives. When an audio CD is played on a home or car CD player, the error correction is always removing errors from the data stream. If an error is not correctable, players will not reproduce the erroneous data. Rather, they will "fill in" the hole left by the missing data with a close approximation of the data. The listener never notices this and everything is fine. When a CD music track is loaded into a computer from a CD-ROM drive, any uncorrectable errors are simply left uncorrected. When a computer file such as a WAV or MP3 is constructed with this data, the errors are treated as legitimate data and left untreated. This results in annoying "clicking" and "popping" when this file played. Any duplicate CD's or MP3's made from the file will have the noises. The record company simply loads the CD with uncorrectable errors, which are smoothed over by CD players, but show up on pirated MP3 or WAV files. This technique is far from perfect since there are software programs that can automatically remove the clicks and pops. Also some marginal CD players may have problems with CD's treated with this scheme.

HIDING THE TOC: Another technique of record companies to foil pirates is to confuse the CD-ROM drive with bogus TOC (table of contents) information. This makes it difficult for the CD-ROM drive to decide what kind of a disk is in the tray. Since run-of-the-mill audio CD players don't have to "decide" what kind of disk is inserted, they just skip over all the erroneous intro information and go directly to the audio tracks. Of course, like the previous technique this is based on a design loophole in the CD-ROM drive and will not work on all models of CD-ROM drive. But it can discourage the casual "Joe Six-Pack" who wants to post a song on a website.

The original audio CD, a joint effort of Phillips and Sony, is one of the most successful products in history, totally replacing the vynl LP within 6 years of its introduction. It is a tough, robust format that can take all kinds of abuse and still deliver flawless audio, which testifies to the cleverness of its designers. The CD begat several formats:

CD-ROM—this is the familiar CD that is used for general purpose computer file storage and software distribution. In making CD-ROM's, an extra layer of error correction is added to the normal Reed-Solomon for increased reliability. This accounts for its legendary accuracy and durability. This came out in the late 1980's. The standard file format for CD-ROM is called ISO 9660.

MIXED MODE CD—It is possible to record a CD with both Red Book audio tracks (playable on any CD player) and CD-ROM data (readable on any CD-ROM equipped PC). The easiest way to do this is to simply mix the ROM track with the music tracks. The ROM track MUST be recorded as track 1, with the music tracks following on tracks 2, 3, 4, etc. This works very well since the CD-ROM player only sees the ROM track while the music player can ignore the first track and just play the music. There is one slight problem, if the music listener tries to play track one through their ordinary music CD player, the player will attempt to decode the data as PCM audio and it will belch out loud unholy noises. If the listener has a $1000 stereo and destroys the delicate ribbon tweeters (or damages his hearing) with this pseudo-audio, a lawsuit could be in the making! This type of mixed mode is generally not used because of this "track one problem."

MULTISESSION CD—To avoid the "track one" scenario of the mixed mode CD, the multisession CD is oftem employed. In a CD, it is possible to have two complete "sessions". A "multisession" or "double session" CD is really two CDs in one. The first session is playable by all CD players ever made. Music players see the first session as the ENTIRE disc and are oblivious to any further information on the disc. Most modern CD-ROM drives, however, can see both the first and second sessions. By putting the music tracks in session one and the CD-ROM track safely tucked away in session two where it is invisible to the music player, the disc can function as a normal music disc or CD-ROM disc. Problem solved.

CD-I or CD interactive is a special format designed by Phillips to hold video, audio, and an application program. CD-I players are actually computers that are hard wired to run the program on the CD-I disc. The biggest application of this format has been for education and industrial training. CD-I machines were, in fact, complete computers running the OS-9 RTOS (real time operating system). Unfortunately, the processor used was a Motorola 68000 series which was only running at 15 MHz, a small fraction of the clock speed of modern CPUs. Internal RAM was only one megabyte, and programs depended heavily on CD data, which is read much slower than a PC's hard drive data. Even with these limitations, many CD-I application discs were made and were quite effective. At the time the CD-I was developed, the 1980s, it offered a cheaper alternative to full-blown PC's for running graphical software. When CD-I came out, A full PC would set you back over $1500 (with only MS-DOS and no Windows) and 1.44 MB floppies were the main storage medium. At that time CD-I was an attractive platform, especially for video games. Now with fully loaded PC's going for less than $600, CD-I is pretty much obsolete. One big advantage of the CD-I format is that the OS s hard wired into the box, so it is immune to the crashes often experienced with some PC operating systems. Also it is a simpler device and is more user-friendly than most PCs.

CD-BGM or CD background music is a sub-format of CD-I. BGM discs look just like regular audio or computer CD's, but they hold up to eight hours of mono FM-quality (15 KHz response) audio. BGM discs were invented by Phillips in the 1980s for distributing "functional" music for use in public places like cafeterias, motel lobbies, banks, etc.BGM discs use a pre-MPEG compression technique called ADPCM which uses 4 bit samples instead of the usual 16 bit samples. This multiplies the capacity by 4X. By using mono instead of stereo, the capacity is further increased by 2X, yielding an 8X total increase. The sampling rate is 37.8 KHz instead of 44.1 KHz. All CD-BGM's can be played on all CD-I machines, but no BGM CD's are playable on consumer equipment. This format is fading fast due to the popularity of more modern MPEG-based coding such as MP3 and hard-drive based music systems, which can hold hundreds of hours of music. The new MiniDisc MDLP format, which can squeeze five hours of audio on a MiniDisc might be a contender for background music also.

PHOTO CD—this format was pioneered by Kodak in the early 1990's for storing high quality photographic images. Like CD-BGM, it is a subset of CD-I standards. In the late 1990s the availability of CD burners and JPEG, GIF, MPEG file formats rendered this format less attractive to consumers.

VIDEO CD—the video CD is a special CD that holds 74 minutes of video and compressed digital audio. Video CD had several strikes against it right out of the box. First, the video is VHS-quality MPEG I, which is inferior to all other video disc formats (DVD and the old Pioneer laser discs). Second,the 74 minute time is not long enough for most full length movies (requiring breaking the movie into 2 discs). This format never caught on in the USA, although it did do fairly well in Asia.

CD VIDEO—Not to be confused with Video CD, CD Video or CDV was hybrid between the audio CD (Red Book) and the Pioneer analog video laserdisc format. This format came out in the late 1980's and allowed about 5 minutes of video to be on the same disk with about 20 minutes of regular CD audio tracks. These discs could be played on a regular CD player for listening to the audio tracks or could be popped into a laserdisc machine to watch the video snippet (usually a music video of one of the songs on the disc). It was thought that the "MTV generation" would fall head over heels for this format but there was not much public interest in the USA. The format did have some success in Asia, but then so did Karaoke.

CD+G or CD plus graphics is an interesting technique whereby simple pictures can be imbedded in the subcode channel of a normal music CD. This technique is not used very often. It requires a CD player with a digital output connected to a special decoder.This format never gained much acceptance outside of Japan, where it is used for karaoke machines.

DVD—or digital video disc is the successor to the LD format. It is the size of a CD and resembles a CD. Unlike a CD, DVD uses a special short-wavelength laser and stores eight times more data and makes extensive use of a technique not available to the architects of the audio CD: MPEG compression. Uncompressed digital color NTSC video requires over 100 megabits per second of storage. Obviously this is very impractical! Using MPEG, this number can be brought down to 10 megabits per second or even less. DVD records compressed MPEG video at a rate of around 9 megabits per second and multi-channel surround audio at less than 500 kilobits per second.

DVD is more than a movie medium. It is designed to be totally "backward compatible" with CD's (i.e. you can play audio CD's on all DVD players made). It is expected that at some point in the future that CD player and CD-ROM drive production will cease and ALL pre-recorded entertainment and software will shift to DVD. The DVD may become the universal "do-all" format for business, information systems, and of course, entertainment.

OPTICAL JUKEBOXES—Increasingly, CD and DVD based media are used for "near line" storage of computer data. Near-line data, like on-line data, is available to the user on demand. The main difference is speed. On-line data is available nearly instantly whereas acquiring near-line data can take several minutes. Optical jukeboxes use CD-ROM, DVD-ROM, or DVD-RAM disks. In the case of DVD-RAM, the disks are treated like hard drives (read and write access). Optical jukeboxes can provide several terabytes of data on request and are often used for medical imaging and video on demand applications. Tape-based jukeboxes (called silos) are also available. High capacity tape systems such as IBM 3590, Super DLT, or Sony DTF are generally faster and have higher capacity than optical jukeboxes, but are more expensive and have shorter lifespan than optical media due to head and tape wear.

8

PHOTOGRAPHIC FORMATS

GENERAL BASICS

Although not an "electronic" medium in the strict sense of the term, photographic film often carries light modulated audio and digital information. Also "flying spot" electron beam techniques can be used to record or read images from film.

Photography is the process of chemically recording optical information. The film is coated with a light sensitive "emulsion" that chemically reacts to light rays to record high resolution images. Each "grain" particle represents one picture element (pixel). The lightness and color of this pixel is part of a mosaic of many pixels that together form an image.

OPTICAL SOUNDTRACKS

Soundtracks in movies are usually recorded by amplitude modulation (AM) of a light source. There are two ways to impart AM sound to a film. Density modulation is a technique where the current of a light source is modulated to vary the light output and thus create AM. This modulated light is forced through a narrow aperture slit and excites the emulsion of the film and imparts the audio. On playback the soundtrack is backlit by a similar lamp (running on tightly filtered DC to prevent hum). This light goes through a narrow aperture and is passed through the variably transparent soundtrack where it picks up the original modulation. A photocell or phototransistor then converts the modulated light back to audio.

Another way to lay a soundtrack on film is called variable area modulation. This technique reflects a tiny spot of very bright light on to the film. Small electromagnets are used to "wiggle" the spot, increasing and decreasing the average area of film that is illuminated. This technique is the most common method for imparting an analog soundtrack on film. It can be detected with the exact same apparatus used for variable density, since it is just another type of AM. (The recorders that lay down the variable area tracks use the same mirror-galvometer arrangement used for trans-oceanic telegraph receivers in the 1800's!) Variable area is preferred over variable density because it has a better signal to noise ratio. Density modulation is more prone to "grain noise". Unfortunately, variable area has much more distortion since it relies on a mechanical movement similar to a phono cartridge.

Modern optical soundtracks often use Dolby (tm) noise reduction for better results. Also there are now systems that record digitized CD-style audio tracks on to the film along with the standard AM optical track(s). Some movies use a totally separate, but synchronized tape or CD-ROM for the sound.

There are also magnetic sound-on-film systems, these impart the sound on a stripe of magnetic tape that is fixed to the edge of the film. These operate in the same way as stand-alone tape recorders. Magnetic sound has fallen into disfavor since adding the magnetic stripe increases the cost of the film.

ELECTRON BEAM RECORDING

A fascinating hybrid between electronic and optical recording is the electron beam recorder. It is possible to use an electron beam to record a video frame directly to photographic film. This technique is often used in Hollywood to transfer digital video generated on computers to 35 millimeter film. Because it is quite expensive, it is limited only to short scenes involving dazzling special effects.

It is also possible to do this in reverse, converting a film image back to video. Many television programs are still filmed on 35 or 16 millimeter motion picture film. In the post-production process, the film is run through a device called a "flying spot scanner". This is a tiny high resolution CRT with a blank video raster. This blank raster is fed through the film and becomes "modulated" by the image on the film (the modulation is picked up via a phototransistor). This modulated image is merged with the original synchronization of the raster. This results in a high quality video signal from film. This video signal can then be transfered to video tape for editing.

In the 1950s and 1960s it was quite common for TV broadcasters to record TV shows on film for delayed playback. This was called a "kinescope" and was the only practical way to "time-shift" programs until low cost VTRs and VCRs arrived.

Electron beam recording has been used to impart very small images on motion picture film and microfilm. In the pre-PC and pre-CDROM era, microfilm was often used to distribute vast quantities of data in a physically small package.

One of the first consumer video playback devices was based on electron beam recording and "flying spot" playback. It was the CBS EVR format (Electronic Video Recording). The EVR Teleplayer was introduced around 1970 by CBS Labs. It was truly an oddity. It used 8 millimeter un-sprocketed film in a "cassette" that was a round reel about 7 inches across. The "tape" would be pulled out of the reel and automatically threaded. Even more strange—the film had two parallel tracks of programming, each 25 minutes in length for black and white (for 50 minutes total). For color, both tracks were used simultaneously for a 25 minute program. During the middle of a black and white program, the user had to stop the film and rewind it to play the second track. This format never caught on with consumers. Its main use was in education and industrial training.

9

SOLID STATE MEMORY FORMATS

With advances in very large scale integration (VLSI) chips and manufacturing techniques, silicon is now used as a storage device. The first solid state transistors were made from germanium, which is easy to work with compared to silicon, but has many disadvantages such as high leakage and high temperature sensitivity. In the late 1950's, Silicon became the dominant material for solid state transistors and allowed for the development of the integrated circuit, which combined many transistors together. Today practically all high density electronics is silicon based.

STATIC RAM—The first type of solid state storage was called a static ram. Static RAM (random acess memory) uses a pair of transistors (called a flip-flop) for each bit of storage. This type of storage, while reliable and very fast, consumes a lot of power. It is therefore mainly used where speed is most important such as internal "scratchpad" caches and for CPU registers.

DYNAMIC RAM—The dynamic random access memory is the main "work-horse" for data buffering in computers. Often called D-RAM, the dynamic RAM uses an array of capacitors for bit storage. A capacitor stores a small electrical charge for a short time. After this time the charge will "bleed off" of the capacitor and the information will be lost. To prevent this, the capacitor charge needs to be refreshed at regular intervals. To do this the "satus" or value of the bit (1 or 0) is sensed and reinforced periodically. While this may seem to be a precarious way of storing data, in practice it works very well and allows hundreds of millions of bits per chip. D-RAM is, of course, used in personal computers for creating, editing, and manipulating data. In recent years, D-RAM has become very inexpensive and is now used in many non-computer devices for storage of digitized audio (like MP3 players and answering machines) and for digitized images (such as digital cameras). D-RAM and static RAM are both "volatile" in that they lose their information as soon as the power is turned off. To keep the information from vanishing, some D-RAM's use a small "button" lithium-ion battery to retain information even when the main power is off. D-RAM recording and playback uses no moving parts, which makes it ideal for "endless loop" playback devices since it does not wear out. D-RAM will never be competitive with magnetic and optical storage for long term reliability and cost per bit, but for mechanical stability and portability it reigns supreme.

In the days before cheap silicon chips, D-RAM was actually implemented using a cathode ray tube (CRT) similar to a TV picture tube. The face of the tube was treated as a storage buffer that would be refreshed by a feedback loop. In fact, the good ole CRT can also be used as a ROM as well. The CRT and other "flying spot" tubes were the first all-electronic mass storage devices. Also pulse delay devices such as shift registers and analog delay lines can be used as "low tech" versions of the D-RAM, where the data is recirculated and regenerated at regular intervals.

NVRAM—NV RAM or non-volatile RAM is an ordinary RAM chip that has a tiny battery to keep the data refreshed when the main power is turned off.

ROM—ROM or read only memory works exactly like RAM except that the information in the chip is permanently recorded in the chip and cannot be changed. Most ROMs are simply a matrix of bits that are "hard wired" to a logical one or zero status. ROMs are generally used to store low level machine language programs in computers. The BIOS or basic input/output system is a set of routines that are used to prepare a PC to run its main operating system (Windows, Linux, DOS etc) It is a ROM chip. ROM chips are commonly employed as "look up" tables for generating characters (such as LED/LCD numerical characters or ASCII characters).

PROM—The PROM or programmable read only memory is analogous to a CD-R. It can be written only once. Once written it then becomes a ROM and can be treated as such.

EPROM—The EPROM is an Erasable Programmable Read Only Memory. It operates exactly like the ROM except that it can be erased and re-written. In this respect it resembles a RAM. The main difference between EPROM and RAM is that a RAM can be selectively erased while an EPROM must be completely wiped clean. Another difference is that EPROMS are not volatile and they retain their data even with no power applied. Also EPROMS must have their information "burned" into them like a CD-R or CD-RW with a special piece of hardware. EPROMS are used for semi-permanent data such as look-up tables. EPROMS are not well suited for general storage as they have low capacity, a high cost per bit, and are not "user friendly".

EEPROM—The EEPROM or Electrically Erasable Programmable Read Only Memory is identical to the EPROM except that it has the ability to erase itself and program itself without a special external device. In this respect it is similar to RAM. However, EEPROMS must be erased FULLY or not at all whereas a RAM can be selectively erased and re-written. Typical use of EEPROM is to store preferences and settings for a computer's BIOS.

FLASH MEMORY—Flash memory is a small cartridge containing non-volatile RAM. This is designed to be treated like a diskette, cassette, or other removable cartridge medium.

ANALOG EPROM/EEPROM—While nearly all EPROM and EEPROM chips store logical one or zero in each bit, it is possible to store analog voltage values directly in each bit of an EPROM or EEPROM. This technique can be used to implement a "tape recorder on a chip." Such a device can take an variable voltage such as audio and store voltage samples directly as analog bits. This allows for a solid state audio recorder without analog to digital conversion. This technique is sometimes employed in little 10 or 20 second "memo" recorders which can be clipped to a keychain and run on a tiny battery. This device has found its way into toys, answering machines, and endless-loop advertising displays.

10

NETWORKING TRENDS

Networking greatly extends the functionality of electronic storage. Digital computer networks allow individuals and organizations to make the most of their storage technology investment. There are several types of computer networks. These can be used for the usual inter-computer purposes such as e-mail and file transfer. They can also be used to carry streaming multimedia content such as audio and video services.

The line between communication and storage has been blurred in recent years because of the rise of multipurpose networks such as the Internet in which storage of content is decentralized and shared by millions of servers widely geographically separated. Even in the arena of personal entertainment electronics, the Internet techniques of packet switching, streaming, and hard drive recording are gradually replacing the old analog systems. For example, DBS satellite services use packet switching and multiplexing similar to the Internet. Also many DBS and digital cable boxes now include a local hard drive "cache" for saving audio and video content as an MPEG (compressed data) file.

The transmission techniques of TV and radio broadcasting are now shifting to digital delivery. This means that radios and TVs will include modems (to convert the modulated radio frequency to direct data), codecs (to decode the data and convert it back to audio or video), and a local cache buffer for temporary storage of content.

TYPES OF NETWORKS

There are basically two broad categories of computer networks. The LAN or local area network connects 2 or more PC's together within the same building, worksite, or campus. The WAN or wide area network connects computers on a regional, national, or international level. The MAN or metro area network is basically another kind of WAN.

LAN DELIVERY SYSTEMS

* ETHERNET—Ethernet is the most popular short distance computer networking protocol. There are two types of Ethernet. The 10baseT uses ordinary twisted pair phone wire to carry up to 10 megabits per second from one PC to another. 100baseT, often called Fast Ethernet, uses twisted pair wire to move 100 megabits per second between PC's. Ethernet is a very popular LAN protocol, thanks to cheap network cards and the use of low-cost telephone wire.

* USB—USB, the universal serial bus, has become extremely popular for connecting computer peripherals such as scanners, printers, and disk drives to personal computers. The USB can carry data at up to 12 megabits per second for distances of up to 15 feet. Repeaters are available to increase this range. Although USB is not a full fledged LAN, it is commonly used for linking and sharing peripherals. One USB hub can address up to 127 attached devices. USB uses low cost twisted pair wire similar to Ethernet.

* FIREWIRE—Firewire, also called IEEE 1394 is not only a computer hardware connection protocol, but a general purpose high speed link to carry digital video and audio using twisted pair wire. The Firewire interface can carry over 100 megabits per second. It is used for computer to peripheral connection as well as linking digital entertainment devices such as set-top boxes and digital video recorders. Range is around 15 feet without repeaters.

* FIBRE CHANNEL—Fibre Channel is a very high speed corporate networking standard. It is designed to connect high capacity tape and disk drives for on-line and near-line storage, backup, and archiving of data. As the name suggests, the main physical medium is fiber optics (single or multi-mode) and speeds in the hundreds of megabits per second are typical. Range is several miles without repeaters. Fibre Channel is not cheap, and is therefore limited to deep-pocketed users such as corporate data centers. It has also become popular in TV studios for connecting video servers and workstations.

* RS232—The RS232 interface can be thought of as a minimal LAN. This is a serial computer-to-computer or peripheral-to-computer interchange cable that

allows for low to moderate speed data (generally less than 100 kilobits per second). It's a fixture on all PC's and is supported by all operating systems. The range of RS232 is 50 feet without repeaters, much more than the SCSI, USB, or Centronics (parallel) maximum unrepeatered distance. The main drawback is that it can only "talk" to one device (one modem, one printer, one other PC, et cetera) whereas the SCSI, USB, and parallel ports can be "daisy chained" with several devices piggybacking on the same cable. Still, RS232 is far from obsolete since it is so easy to implement, using as few as three conductors for duplex communication.

* SCSI—The Small Computer System Inferface, called "scuzzy" can also be thought of as a minimal LAN. Originally made for connecting external disk drives to computers, it has now evolved into a general purpose peripheral connection protocol. SCSI has a couple of downsides. It requires a fairly bulky heavy cable and has a range of only a few feet. The upside is speed. It is very fast compared to other interfaces.

* CENTRONICS PARALLEL—The centronics interface is often referred to as "the parallel port". It is mainly used for connecting PC's to printers. These days, most printers use USB instead of parallel, so this format is not used as much. Besides printers, some external disk drives such as Iomega Zip can use this bus for data interchange. The range of the Centronics port is about 15 feet, but repeaters can extend this to hundreds of feet if necessary.

* Wireless LAN's—There are several competing formats for wireless LAN's. Generally these use 900 MHz or 2400 MHz bands (which are set aside for non-licensed wireless consumer devices). One drawback to these systems is that they use frequency bands that are becoming increasingly crowded with such diverse users as microwave ovens, baby monitors, cordless phones, security cameras, remote VCR transmitters, and even department store theft prevention tags! In urban areas, these bands are quite congested. Fortunately, wireless networks (along with most cordless phones) use a technique called spread spectrum or CDMA (code division multiple access). This spreads the signal out across the entire band and requires a similar de-spreading code at the receiver. This allows many users to share the same band without noticable crosstalk or interference. Another nice benefit of this approach is that it is almost impossible for an eavesdropper to decode it, assuring nearly "wire grade" privacy.

WAN DELIVERY SYSTEMS

The wide area network is used by corporations and organizations for sharing of files and delivery of real-time audio and video. Generally WAN's are carried over leased telecommunication facilities such as satellite channels and fiber optic facilities.

* ISDN—Integrated services digital network. This is a switched digital connection, which means that the user must "dial" another user to establish a temporary communication circuit. ISDN connections are temporary, like ordinary voice phone calls, making this option only viable for occassional access instead of continuous access. Bit rates for BRI (basic rate ISDN) are 64-128 kilobits per second. PRI or primary rate ISDN allows up to 1.544 megabits per second transfer bandwidth.

* T1 or DS1—This is a permanent (always connected) fixed rate circuit that allows up to 1.544 megabit per second of constant transfer between two points.

* T3 or DS3—This is a permanent (always connected) fixed rate circuit that allows up to 45 megabits per second of constant transfer between two points.

* FRAME RELAY—This is a variable bandwidth circuit between two points. This service uses variable length packets and statistical multiplexing. A minimum and maximum transfer speed is guaranteed, but the throughput varies within this range. This is much cheaper than dedicated T1 or T3 bandwidth for comparable throughput, but it is not good for constant bit rate (CBR) data such as streamed audio or video. It is an excellent choice for non-realtime data transfer such as FTP (file transfer protocol) or e-mail. Like T1 and T3, it is always connected. Frame relay is a packet switched network (like the Internet) as opposed to ISDN and regular voice phone, which are circuit switched networks. Frame relay is generally used for linking WAN's and for Internet access.

* ATM—ATM or asynchronous transfer mode is a packet switching protocol similar to frame relay except that the packets are all fixed length and can be prioritized for different QOS (quality of service) levels. ATM can be set for varying levels of throughput, allowing delays for non-realtime data to take advantage of statistical multiplexing while at the same time preventing delay for CBR (constant bit rate) packets. This makes ATM ideal for carrying streaming audio or

video channels as well as for data transfer. ATM, unlike other packet protocols, allows for a fixed propagation path to be established between two (or more) points in the network. This contrasts sharply with IP (Internet Protocol) which simply shuffles packets blindly from one URL (address) to another using a variety of routes. Although ATM is a packet switched system, it is often used to emulate circuits such as T1 or T3. Think of it as "smart" TDM, as opposed to SONET, which can be thought of as "dumb" since it simply assigns time slots without regard to the nature of the data.

* X.25—X.25 is an old (1970's) packet switching protocol that was designed with aggressive error correction. It was designed to run on noisy copper wires which required this extra error handling system. It has fallen out of favor today since the latest fiber optic networks have far less noise than the old copper trunk lines and all the extra "overhead" inherent to X.25 is not necessary. Frame relay has largely replaced X.25 for today's more robust data networks.

* VSAT—VSAT or very small apeture termninal is a satellite based WAN technology. VSAT's are generally used in a point to multipoint configuration. The central hub, which is usually at the corporate headquarters. is able to transmit data (to the satellite transponder) at rates of several megabits per second to all of the connected "spoke" stations. The spoke stations are allocated bandwidth on a rotating basis (like token passing). Each spoke station waits for the opportunity to transmit. When permission to transmit is granted, it can send a "burst" of data during its allocated time slot. If it does not finish during this time, it must stop and wait for the next time slot to finish. This is somewhat like a Token Ring LAN. VSATs generally operate on the KU satellite band (10-15 GHz) using SCPC (single channel per carrier) narroband channels. The VSAT hub uses a large high gain dish with hundreds of watts of power. Its outbound data rate is 1-5 megabits per second. The spoke stations receive data (which can be addressed to any or all spokes) using small dishes the size of mini satellite TV dishes. They send data back to the hub using much lower power (generally about 10 watts) at a much lower data rate (less than 100 kbps). The VSAT is ideal for large retail chains for inventory management,credit card authorizations, and even one-way streaming audio such as in-store advertising.

* DSL—DSL or digital subscriber line is a relative newcomer to the WAN world. DSL is normally used to connect homes and businesses to the Internet, but it can also be used to connect to a private WAN instead of the Internet, (The Internet is

just a large public WAN!) DSL uses frequency multiplexing to permit high speed data and normal voice phone calls on the same wire pair. To accomplish this, a frequency splitter is employed that allows the band from 0-5 KHz to pass to the phone without alteration. The band from 5 Khz to about 1 MHz is used to carry high speed 2-way data. Modulation methods vary, but DMT (discrete multi-tone) and CAP (carrierless amplitude and phase) are two popular systems. There are several sub-formats of DSL, but the most common of these is ADSL or asymmetrical DSL. In this approach, there is more bandwidth on the downstream side (from the phone company) than from the subscriber. Bit-rates are typically 500 kilobits to 2 megabits per second on the downstream (phone company to subscriber) side and 100-500 kilobits per second on the upstream side.

* DVB—DVB or digital video broadcast is a global set of standards for the one-way transmission of digitized video and audio via terrestrial transmitters, cable, and satellite. DVB is a unidirectional WAN technology. It uses packet switching and multiplexing similar to ATM. DVB is used for direct broadcast material (i.e. Dish Network) and non-broadcast private content such as training videos and in-store retail advertising. Each "channel" in DVB is represented by a packet stream. These packets contain header information, raw data, and redundant error correction data. DVB packet streams are typically woven together using statistical multiplexing. The aggregate packet stream is then modulated to a satellite transponder, cable channel, or broadcast channel. The video carried by DVB is usually MPEG-2 and the audio is usually MPEG-1 Layer II. The DVB standards are being adopted all around the world for direct broadcast satellite, terrestrial over-the-air transmission, and cable distribution of audio and television material. DVB can operate in an encrypted mode (called conditional access) to limit reception to authorized customers or in the FTA or "free to air" mode which is unencrypted.

Unfortunately, DVB has not been fully embraced in the good ole USA. In America, proprietary standards from heavyweights such as Motorola and Zenith rule the roost. There are several "DVB-like" proprietary standards such as the Motorola Digi-Cipher II, used for distributing TV via satellite and the Hughes DSS system used for the DirecTV service. These are very similar to DVB but deviate somewhat in certain areas such as packet size. In the terrestrial area, again it's a proprietary standard, the Zenith 8-VSB system, that America has adopted. This is basically the same as the DigiCipher II, except that the modulation is discrete level AM instead of PSK (phase shift keying) used for the DigiCipher II. In contrast, Europe has fully embraced DVB protocols, such as DVB-T for terrestrial broadcasting, DVB-C for cable, and DVB-S for satellite. Most of Asia and Africa will eventually go DVB for their TV and audio distribution, leaving the United States as an island of incompatible standards. There are some exceptions to this trend such as Dish Network (owned by Echostar) which is compliant with DVB-S and some cable systems implementing DVB-C compliant set-top boxes.

* IP—IP or Internet protocol is the most well known packet switching scheme. It was developed by military researchers in the 1960's and 1970's. Although it is old, the popularity of the Internet and Internet client and server software has guaranteed that this protocol will be around for a long time. IP is implemented on the Internet of course, but also on many private "closed circuit" networks such as corporate WAN's and LAN's. The advantage is the ability to use well known and inexpensive web browsers and servers such as APACHE (which costs $0) that many people are comfortable with. Often, IP is run atop of ATM or frame relay. The disadvantage of IP is that it has cumbersome packets with variable propagation delay that makes it a bad choice for streaming media. This can be mitigated by careful network planning.

* SONET—SONET is an acronym for synchronous optical network. This is a straight TDM (time division multiplex) hierarchy that combines digitized phone traffic (normal voice telephone service) plus digitized video and audio, plus computer data together. The smallest channel handled by SONET is called OC-3. It is somewhat bigger than three T3 circuits (155 Mb/s) and is designed to handle 3 T3's plus extra "overhead" data for routing. Alternately, it can be treated as one large channel with "concantenated" bandwidth (called a OC-3c). OC-3c is often used as a pipe for handling ATM or IP for WAN's or the public Internet. Long haul SONET systems do not manipulate anything below the OC-3 level (such as T1). Thus SONET is a "wholesale" level system that focuses on hauling large packages of data similar to the way an 18-wheel freight truck handles mainly large shipping cartons. Along the route of a SONET system, there are add-drop multiplexers that remove data from the system (to be routed to a smaller local network) and add new data to the payload. SONET is synchronous, meaning that all the channels (called pipes) have fixed bandwidth with a recurring time slot, which is inefficient for "bursty" traffic like computer file transfers since the time slot occurs even if it is not needed. To improve the efficiency of SONET, it is often coupled to asynchronous protocols such as ATM, frame relay, and IP that allow for statistical (demand based) time division multiplexing. SONET is a circuit

switched system that creates temporary and permanent fixed-bandwidth paths between two or more locations. Most of the regional, national, and international telephone traffic, plus the major Internet backbone routes are carried via SONET on fiber optics.

ANALOG DELIVERY SYSTEMS

Although digital techniques for information delivery are the most cost effective for most applications, analog systems are still viable for short distances. In the telephone, cable TV, and terrestrial radio industries analog and digital will coexist for many years to come.

* <u>LOCAL TELEPHONE SERVICE</u>—Telephone service is carried by a mixture of analog and digital technology. All long distance and most intracity telephone calls are handled through digital fiber optics running SONET. (Some corporate telephone traffic is running over WAN protocols such as IP or frame relay). On the "wholesale" level all phone service is digital, The only analog aspect of the modern telephone is the "last mile" which is called the local loop. The customer phone is connected to the phone company through a wire pair. Current to run the phone is provided through this loop. This loop carries a bidirectional audio channel to and from the subscriber. In days of old, the phone set consisted of a carbon microphone (which modulates loop current with voice frequencies) a high impedance earphone, and a special transformer called a hybrid that allows bidirectional operation. Today phones have condenser microphones and use electronic amplifiers, but they function the same way. The phone channel is designed to carry voice frequencies from 300-3000 Hertz to and from the subscriber.

* <u>ANALOG PHONE LINES</u>—It is possible to lease a twisted pair phone circuit from the phone company that does not go through the switch (cross connect) apparatus. Such a line can be "equalized" to handle a range of audio frequencies. An equalizer gives the line a flat bandpass within a desired frequency range. This type of circuit can be used to carry a closed circuit audio program, telemetry (i.e. burglar alarms), or data. This is called "dry" wire.

* <u>COAXIAL CABLE</u>—Coaxial cable is used for hauling television channels for short distances (1000 feet or less). Although many cable TV companies carry digital channels, cable TV is still mostly analog. Traditional analog cable TV head-ends generate a batch of side-by-side VSB-AM (vestigial sideband which is a

narrow band form of amplitude modulation) channels that can be directly viewed on a TV as if they were received by an antenna. A coaxial cable typically carries a bandwidth of up to 1000 MHz. In most cable systems, the signals are first modulated to fiber optics, which can go for tens of miles without repeaters. The actual subscriber cable (the so-called last mile) is the co-ax cable. Another use for co-ax is in-building and inter-building closed circuit systems for TV, audio, and data. Video on a coaxial cable can be sent modulated to an RF (radio frequency) channel (identical to an on-air TV channel) or sent baseband (directly without an RF carrier).

* ANALOG SATELLITE—Satellites are orbiting repeaters that take a band of RF (radio frequency) from the ground, amplify it, and redirect it back to earth. This RF can be analog or digital. Satellites have very wide bandwidth (hundreds of megahertz) but are very noisy. For this reason, FM (frequency modulation) is employed exclusively for analog satellite transmission. The FM can be modulated with video or audio signals. In the past few years, much of the traffic on communication satellites has been shifting to digital modulation which has better quality, security, spectrum efficiency and reliability than analog. Because of the high noise level of satellites compared to terrestrial systems, special noise reduction techniques are necessary for any analog signal.

* ANALOG FIBER OPTICS—Fiber can handle analog or digital signals. The advantage to analog on fiber is that the terminal equipment is much simpler and cheaper than digital terminal equipment. A fiber is a waveguide for light. This light is generated by an L.E.D. (light emitting diode) or a small laser. Engineers describe fiber optics as "quiet" and "fast". This means that there is very little noise and a very high bandwidth. It also means that fiber can carry enormous volumes of information for tens of miles without any repeaters, in sharp contrast to copper cables, which need repeaters spaced at intervals ranging from a few feet to a mile (depending on bandwidth). This is one reason why cable TV companies converted their intermediate "trunk" lines from heavily repeated copper coaxial cable to fiber optics.

* BROADCASTING—Television and radio broadcasting still employs analog RF (amplitude and frequency modulation of a radio frequency carrier). Television is in the process of converting to a digital system, which will make analog sets useless without a special decoder box, Radio broadcasting is mostly analog at present, A new DAB (digital audio broadcast) system has been approved that will

allow AM and FM stations to send a digitized version of their current offerings. DAB will eventually replace AM and FM. Digital satellite and cable broadcasting is already a reality. Invariably all video and audio digital broadcasting uses MPEG-based technology to compress the audio and video data to a reasonable bandwidth.

The new digital TV system (in the United States) for over-the-air digital TV is called 8-VSB. It is an AM-based digital system. Unlike analog AM, where the carrier is variably modulated, this new system sends out eight possible amplitude values. At any given time the transmitter is sending one of the eight values. This allows for each amplitude value to represent three bits of data. With its six megahertz wide channel, a TV station can send over 19 megabits per second of broadband data. This is enough for one high definition channel or up to four standard definition channels. This also allows TV stations to become "common carriers", leasing out their extra bandwidth for non-TV purposes such as paging or software distribution. Recently a new digital radio standard called "IBOC" for in-band on-channel was approved in the U.S. This allows a radio station to piggyback a digitized version of its audio feed on the same carrier used for analog. This system appears to have been "railroaded" through the approval process without much forethought and has many unresolved interference issues.

11

THE FUTURE

IMPACT OF MOORE'S LAW

There is no way to know what formats will predominate in 10 years in electronic media storage, but we can expect that many current trends will continue. Moore's law, first put forth by Intel cofounder Gordon Moore in the 1960's seems to be holding up. Briefly, this concept is that the number of integrated transistors, and thus the computational "muscle" of computer hardware, doubles at approximately 18 month intervals. From the early 1970's to the present (2002) this has been surprisingly accurate. We are pushing the limits of silicon based integrated circuits by squeezing ever increasing numbers of transistors, resistors, capacitors together. The dominant high-level integration technique today in computers and most high-density electronics is called CMOS or complementary metal oxide semiconductor. The basic building block of CMOS is the MOSFET or metal oxide semiconductor field effect transistor. By scaling MOSFETS down smaller and smaller (and the associated passive components) we can achieve faster microprocessors and higher capacity RAM and ROM storage. This is having a profound impact on not just computers, but many information appliances such as TV set-top appliances, audio devices like MP3 players, DVD players, etc. Having faster and more powerful microprocessors allows for more powerful signal processing (i.e. MPEG). Having lower cost RAM will allow small NMP (no moving parts) playback devices for audio and video.

Although Moore's law has served us well so far, it may run into a brick wall in the next few years. At some point we will reach the hard limit for further integration of CMOS. When this happens, we will have to look to alternatives for the next level of "smaller-faster-cheaper" electronics. It may be that the basic building blocks of life—organic molecules—which are already used in CD-R and DVD-R dyes, may hold the key. Other tantalizing technologies include quantum computers, single molecule transistors, and 3-D crystal holography. Who would have thought 100 years ago that today's information technology would be based on melted sand?

EMERGING TECHNOLOGIES

There are many exciting new technologies to watch in the information storage and transmission field. In the past twenty years we have seen hard drives go from the dollar per megabyte level down to nearly a dollar per gigabyte. This is having an enormous impact not just on the world of information management, but on the entertainment field. The hard disk is showing up in set-top television boxes, stereo components, cameras, and even wristwatches. It is transforming the entertainment landscape from a top-down centralized landscape to a peer-to-peer file-sharing landscape. Many companies are trying to fight this trend, but this is the future. The down side of this is that it is getting more difficult to control access to entertainment (and information) products and services. This "peer to peer" model is already undermining the recording industry, the publishing industry, and Hollywood, all of which are operating under a twentieth century model of centralized control and top-down distribution. This old business model will have to be revised for this brave new world.

Another trend to watch is packet-switching. Today millions of people use the Internet, the most obvious application of packet-switching technology. Packet switching is replacing the older model of "circuit switching". Briefly, circuit switching is like your local phone company. When you place a phone call you dial a number. This automatically sets up a temporary two-way voice channel for the duration of the phone conversation. When you hang up, the circuit is dissolved. In packet switching, the user sends out packets from one address to another. These packets are little pieces of data. They can represent an E-mail, a photograph, and audio clip, a movie, a song, or any other form of information. These packets are shuffled from your address to that of the other user or users. Each packet has an address imbedded in it that tells the network where it belongs. These packets can also contain real-time digitized voice or video. The main advantage of this technique is that it makes much better use of network resources since bandwidth is only allocated as needed. Packet switching is not limited to the Internet. Digitized TV channels on cable and satellite are sent as packets. The information recorded on DVD's and CD's are a form of packet switching.

The final trend to watch is CDMA or Code Division Multiple Access. CDMA is going to make the RF spectrum an unlimited resource and reduce or even eliminate the need for the FCC (Federal Communications Commission). The old twentieth century model of radio spectrum management is based on analog transmission modes such as AM (amplitude modulation) and FM (frequency modulation). These modes of transmission are a form of FDMA (Frequency Division Multiple Access) whereby stations in the same area are separated by their assigned frequency channel. With CDMA, each channel is mixed with a noise-like "spreading" code on the transmit side. This causes the channel to spread out over a wide area of RF bandwidth. In the receiver, a "de-spreading" code is used to recover the original signal. Thus channels are not separated by frequency, but by spreading code. This permits a great deal of frequency overlap with no interference. It also permits nearly unlimited re-use of radio frequencies. Since the main mission of the FCC is handing out radio frequencies under the old FDMA model, they become obsolete along with the centrallized corporate domination of the radio bands. CDMA is already being used in mobile telephones and other low power point-to-point applications. It is an advanced modulation system that requires fast, powerful logic circuits. But that is no problem these days. (Your DVD player is no less complex and yet is a mass-market item). Eventually CDMA will result in true "any-to-any" conductivity and make the Internet truly universal. You will not "log on" to the Internet. You will wear the Internet. Of course this will raise new privacy issues, but lets not open that can of worms yet!

0-595-30939-9